T0068138

CLIMATE FETISH

Previous Books

"Climate of Ecopolitics – A Citizen's Guide" 2008 iUniverse, Inc.

"Green Gone Wrong - Ecopolitics Exposed" 2001 Writers Club Press

CLIMATE FETISH

Ecopolitics

Paul Taylor

CLIMATE FETISH
ECOPOLITICS

iUniverse books may be ordered through booksellers or by contacting:

iUniverse
1663 Liberty Drive
Bloomington, IN 47403
www.iuniverse.com
844-349-9409

ISBN: 978-1-6632-5729-1 (sc)
ISBN: 978-1-6632-5730-7 (e)

Library of Congress Control Number: 2023920302

Print information available on the last page.

iUniverse rev. date: 10/24/2023

Dedication

This non-fiction book is dedicated to all of the brilliant, brave and diligent global independent climate scientists consulted and referenced herein, and who unfortunately work in a hostile and shamefully partisan political atmosphere.

Contents

CONTENTS

Preface

I have committed my life to understanding and communicating the vast scientific, social, political and economic impacts of environmental problems, and their solutions.

This, my third book is a work of non-fiction is an overview assessment of 21st Century environmentalism where *ecopolitics* have replaced rational scientific discovery and discourse in the hysterics surrounding the purported *existential threat* theories of global climate change. Global warming has been identified as both the world's greatest crisis, and the world's greatest hoax – neither is true.

This book's title use of the word *Fetish* is not meant in the modern vernacular of implicit *carnal* behavior. Rather, *Fetish* implies the original meaning: *an object of irrational reverence or obsessive devotion, a rite or cult of worshippers.*

The term *environment* essentially means *surroundings.* Unfortunately, its meaning has been stretched, conflated, contorted, misappropriated, abused, bastardized and politicized to the point of trivia by activists and media to manipulate public policy in a range of issues that are as endless in their scope as often misguided in their ends. Environmentalists have become skilled at gaming the government regulatory systems for political advantage in the guise of progressive public service as thousands of tax-exempt nonprofit organizations. This is today's *ecopolitics*.

Acknowledgement

The publication of this book has reasonably applied *best efforts* in citing and referencing sources consulted in its preparation pursuant to the *Fair Use Doctrine*. The author wishes to express his sincere gratitude to the consulted, referenced sources in adding to the body of public knowledge on subject matter as important, dynamic and often intractable as the environment and climate change.

My colleague Environmental Scientists' and university students' curiosity over the last five years of university faculty work have motivated and inspired me to write this my third book.

Introduction

Today, computer models can generate compelling scenarios for any political argument on an untestable proposition about a future hypothetical environmental threat. Cloaked in a veneer of pseudo-science, a hypothesis often sustains positions of environmental activists where a tenuous scheme of worst-case assumptions requires the rational observer to prove the irrational negative proposition. Here, the obscure, often immeasurable environmental impact can be promoted as an imminent existential threat. This *fear mongering* has become the routine and systematic, yet disingenuous tactic to erect counterfeit public issues for global political exploitation.

The reader is invited to consume this work with a mind open to the new idea that global environmentalism has devolved into *ecopolitics* as a partisan political special interest in the 21st Century. Sadly, consensus science that solves real environmental problems has been replaced with pernicious political propaganda and demagoguery. And, the cost-benefit analyses necessary for prioritizing and solving environmental problems goes unmentioned in the issues of global climate change – unmentioned because such analyses are incalculable in today's knowledge of climate science.

This book is a rational synthesis of the massive, and often contradictory, volumes of information on global warming and climate change for citizens consumption. This book cites numerous contemporary and credible experts on the issues of climate science and climate policy for a balanced assessment.

CHAPTER 1

▼

ECOPOLITICS

Ecopolitics is a political ideology that aims to foster an ecologically sustainable society often, but not always, rooted in environmentalism, nonviolence, social justice and grassroots democracy. It began taking shape in the western world in the 1970s; since then green_parties have developed and established themselves in many countries around the globe and have achieved some electoral success.

The political term *green* was used initially in relation to *die Grünen* (German for "the Greens"), a green party formed in the late 1970s. The term political ecology is sometimes used in academic circles, but it has come to represent an interdisciplinary field of study as the academic discipline offers wide-ranging studies integrating ecological social sciences with political economy in topics such as degradation and marginalization, environmental conflict, conservation and control and environmental identities and social movements.

Supporters of the green politics share many ideas with ecological issues, green politics is concerned with civil liberties, social justice, nonviolence, sometimes variants of localism and tends to support social progressivism.

Green party platforms are largely considered *left* in the political spectrum. The green ideology has connections with various other eco-centric political ideologies, including ecofeminism, eco-socialism and green anarchism, but to what extent these can be seen as forms of green politics is a matter of debate. As the left-wing green political philosophy developed, there also came into separate existence opposite movements on the right-wing that include ecological components such as eco-capitalism and green conservatism. (Wikipedia, 2023)

Today, *Ecopolitics* is the intersection of government environmental controls, environmental activism, and your freedoms and prosperity. Ecopolitics can be associated with the radical "Woke" environmental justice movement. News and entertainment mass media have become indistinguishable and readily exploitable as an effective propaganda machine in the game of ecopolitics. In today's ecopolitics, partisan environmentalists convene as media events. Politicians arrive to say "they care" to placate the environmentalists for short-term media and electoral rewards, enabling the irrational government regulatory policies to expand without end or accountability to scientific cause and effect, or economic damage. Green environmental policymakers have little patience for rational scientific evidence.

Radical environmental policy is now achieved through regulatory fiat to calm political activists who themselves can no longer be bothered with letting scientific rigor get in the way of what they want. This is a tragic legacy of the 1960s and 70s cultural revolution; where the "virtue signaling" and well-intentioned compassion-baiting dialogue make truth a negotiable commodity to advance one's public political ideology.

Environmental activism can have its own negative impacts, where the sanctimonious environmentalist's obsessive insistence that you change your lifestyle toward eco-purity begins to impair personal and professional relationships – e.g., *climate shamming*. Cultish and fetishistic political devotions are displayed in radical environmentalism. The enduring negative impact of environmental activism is the constant politicization of your way of life and, ultimately, limits to your comfort, freedom and prosperity.

Few, if any, public issues have the personal resonance of environmental issues. Each of us can feel both victim and perpetrator in environmental issues. Environmental issues have become an integral part of world culture, and for some, a personal moral cause or even fetish. The environmental movement has grown worldwide to become the largest, most densely organized political cause in human history. And, the global environmental regulatory systems, begun in earnest in the 20th Century, have often erred on the side of caution under the paternalistic maxim of "prudent avoidance" without complete science for the actual cause-and-effect or economic damage in environmental regulations. Every personal and professional activity in your daily life involves some environmental regulation. Now, global warming theorists claim its cause and effect in every other known environmental issue – global warming has become environmentalism's universal scapegoat, obsession, excuse, and even fetish.

Over 600 global maladies have been attributed to global warming including such preposterous things as a Minneapolis bridge collapse, the Katrina disaster, the decline of circumcision, brain size, cremations and earth's rotation. A list of these wildly speculative and often silly items is contained in Chapter 4 herein.

Predictably, environmentalists identify the sinister enemies of the environment as greedy corporations, property owners and the prosperous lifestyles of developed democracies. It starts with media access that promotes civic fear and misapprehension about environmental issues, and points to capitalist democracies as conspirators against you and our environment. Here, environmental activists become vital media assets, perceived as heroic *saviors* of the planet. This is the *Alice-in-Wonderland* area where the possible crisis becomes the probable, the suggestive crisis becomes the conclusive, the accusation of crisis becomes evidence. Unfortunately, the long-term cost-effective solutions to environmental problems derive from moderate centrist policies that do not attract radical activists, crisis-hungry media or political extremes.

From a broader perspective, and without a clear scientific consensus on the validity of such things as global warming, activists have hijacked political agendas in climatic doomsday scenarios that, if true, are the most apocalyptic so far. However, if false, the trendy global warming hysteria may be the final blow in a precipitous slide in legitimacy of costly environmental activism, and reveal *ecopolitics* as corrupt. For example, in the 1960s and 70s, environmentalists successfully defeated the expansion of nuclear power plants in the U.S. that could have substantially reduced greenhouse gas-producing fossil fuel use impacts which those environmentalists now blame for global warming. Nuclear power plants also reduce U.S. dependence upon foreign oil.

Neither politics nor legal systems have been accountable consistently to the truths of environmental science or economic feasibility. Environmental laws and regulations of any kind have the effect of incrementally limiting personal liberty and freedom, and of imposing severe, and often unnecessary, financial burdens on those who can least afford them. The recent progressive "war on carbon" to mitigate global warming now threaten the economies of entire countries.

When dealing with attitudes toward the Earth's ecosystem, we must recognize the absence of a commonly accepted philosophical or economic basis for the development of an environmental law system. And this development of law occurs in a political setting that is often acrimonious, for the basic political divisions in our society are challenged by new laws dealing with the environment. Thus, a clear and wide political divide has developed between climate "Believers" and climate "Sceptics." See Chapter 5 herein for examples.

Nature was once thought to be the enemy against whom man struggled to survive. Later, nature became the source of exploitable natural resources and economic wealth. In recent time we have begun to appreciate and analyze the relationship of man with nature, and man's interdependence with nature. This "newer" theory calls for man to utilize nature's bounty, but to exist so that the dynamic natural system survives, hence *sustainability.*

The historical development of environmental laws and regulations parallels the evolving philosophical views of nature. Prior to the beginning of the 20th Century, the legal system was used to encourage the development and exploitation of our natural environment through land use rights. The government encouraged highways, railroads and canal building, and gave away millions of acres of public lands to those who would exploit them. Beginning in the late 19th Century, the concept of conservation became popular, and this resulted in legislation in the early part of the 20th Century to conserve and manage our natural resources.

Conservation means many things to many people, and shortly after the conservation movement began, it split into two camps or schools of thought – a dichotomy that exists today. John Muir (American naturalist, 1838-1914) led the protectionist or nonexploitation conservationists, while Gifford Pinchott (American forester and politician, 1865-1946) was the leader of the "careful extraction" or "wise-use" school of conservation.

Conservation is a way of life. And, it should be made clear that living in harmony with nature need not require the giving up of many of the advances that a thousand years of applied science and technology provide human civilization. Science has to do with proving physical facts and relationships using the *Scientific Method*. Science requires the exclusion of human emotions and political preferences to discover truth. Science is the antithesis of humanism.

Climate change polarization research often draws upon social identity frameworks, wherein partisanship is conceptualized as a group identity or social affiliation. People are assumed to be socialized into a partisan identity at a young age, which often remains stable over their life course. One major implication of adopting an identity framework is that many partisans will shift their attitudes (and possibly behaviors) to match what they consider to be appropriate for their social in-group. This often occurs via a cue-taking process, wherein partisan elites (for example politicians, media figures) frame salient issues, which are subsequently adopted by their co-partisans. While some partisans will update their opinions rather abruptly when presented with an elite cue, partisan responses to elite cues are not uniform. The social identity perspective on partisanship helps explain why there is intense polarization surrounding high-profile issues: partisans change what they think to match their group, creating a uniformity of opinion within a partisan social group. (www.Nature.com, 2023)

Have no doubt that environmental awareness has benefitted both the human and natural environment with science-based regulatory controls. However, eco-activism has also spawned an "axis of antagonism" and chorus of *climate shamming* in militant and litigious eco-groups that are an *echo chamber* of fear-mongering propaganda the world over. *Climate shamming* has become a common manipulation effort of green activism. Expanded science literacy could mitigate the hysteria of climate change and other environmental issues, and vastly improve science-based environmental policymaking. We trust that this book contributes to environmental literacy.

Ecopolitics is the intersection of environmental regulations, environmental activism, and your freedom and prosperity.

This book exposes the fear mongering, climate shamming, carbon demonization, regulatory tyranny, green fetishes, nonprofit (e.g. non-government organizations "NGOs") excesses, energy crises, inflationary costs, grifters and enviro-industrial complex corruptions that characterize todays Ecopolitics. This book also supports greater environmental science literacy instruction as requisite university curricula.

CHAPTER 2

▼

ENVIRONMENTAL GOVERNANCE

The environmental movement emerged in the early 20th Century as a collection of conservationists, naturalists and bird watchers. The movement grew slowly until government began recording natural resource impacts of human activity, such as over hunting, timber clear cuts and strip mining. In the last half of the 20th Century the movement exploded worldwide on government recognition of the potential health effects of environmental transgressions. Today, one is considered to be uncivilized if unconcerned about the environment. Today, environmental matters are a free-for-all of global political pomposity, climate anxiety and propaganda.

An example of conservationism was the Rivers and Harbors Act of 1899. This law was enacted to maintain the navigability (transportation) of U.S. waterways. It later embraced aquatic habitats such as wetlands, vernal pools and riparian resources as protected "waters of the U.S.", and has long been the subject of controversy even up to the Supreme Court in 2022.

It has been said that the original environmentalist was the U.S.'s 26th President, Teddy Roosevelt, who established the National Park Service in the early 1900s to set aside wilderness lands and conservation areas. During his 1901 to 1909 Republican presidency, Roosevelt designated 150 national forests, the first 51 federal bird sanctuaries, 5 national parks, the first 18 national monuments, the first 4 national game preserves and the first 21 land reclamation projects. He placed 230 million acres of land under federal protection. Teddy Roosevelt was the first to use the word "conservation" in describing U.S. government policy. Roosevelt had come to the view that "the irresponsible use of natural resources is a fundamental problem which underlies almost every other problem of our national life."

With two World Wars and The Depression intervening, it would take nearly sixty years to complete a federal government framework for conservation. Initially, the concern was smoke, sewage, and such. The clean air, clean water, and solid waste acts of the 1960s were still animated mainly by aesthetic insults or the appearance of being dirty, offensive or untidy.

U.S. Government had become formally committed to encouraging "productive and enjoyable harmony between man and his environment." This promise was made in the National Environmental Policy Act of 1969 (NEPA). Passage of the NEPA in 1969 was an attempt to create a new frame of reference for the consideration of all major activities by the Federal Government – a frame of reference that would include consideration of impacts to the environment. NEPA was the culmination of a decade of previously unsuccessful Congressional attempts to define and put into practice a national environmental policy.

NEPA had a three-fold purpose: to establish national environmental policy, to authorize research concerning natural resources, and to establish a council of environmental advisors. Later Federal action required agencies to prepare an "environmental impact statement" if any of their proposed actions might significantly impact the environment. Another purpose for enacting NEPA was to fulfill the need for an interdisciplinary approach to environmental management and decision-making in all branches and levels of the Federal Government.

Even as the Federal Government completed the regulatory framework for traditional conservation, these laws also quietly launched the new era of environmentalism to control pollutants. To begin with, regulating pollutants of any kind requires a more elaborate and intrusive regulatory force than regulating the more human-scale parks, wilderness lands and wildlife. In 1970, President Nixon established a new cabinet-level agency under the US Department of the Interior, that would be the U.S. Environmental Protection Agency (EPA). The EPA was established to take charge of all U.S. pollution control and associated regulatory programs. More significantly thereafter, each of the new laws also included something quite new – an open-ended "toxics" provision, a general invitation for the EPA to monitor remote and invisible environments for hazardous pollutants and regulate them as needed for public health and habitat protection. In addition, though written with cougars and wilderness mainly in mind, the Endangered Species Act of 1973 (ESA) had been similarly expanded broadly enough to protect such exotics as the flower-loving fly, the fairy shrimp and other critters whose actual numbers in their natural habitats can never be counted. The ESA would soon be expanded to cover "habitat modifications" as well. Progressively-green California adopted its own California Environmental Quality Act and Endangered Species Act in 1970.

A mere statutory afterthought in the 1960s, remote and invisible environmental pollutants get entire Federal Laws of their own a decade later. The Toxic Substances Control Act is implemented in 1976. Then the Superfund hazardous site cleanup program in 1980, followed by RCRA, CERCLA and SARA and other ubiquitous acronyms for government regulatory expansion to monitor and control every detectable impurity, which technology now allows us to regulate at concentrations of less than one part per billion. You are reminded that the EPA began with a budget of $1 billion and 4,000 employees in 1970. Today, the EPA's budget is about $10 billion with 15,000 employees. This does not include the thousands of contract regulatory researchers operating under the EPA-administered grants whose millions of dollars will never be accounted for. This U.S. growth and enforcement in environmental regulations has largely been replicated in other western democracies until today where the United Nations (U.N.) and European Union (E.U.) have lead the global concern over climate change and global warming. Along with the escalation in environmental laws and regulations, the U.S. also has invented and shared numerous pollution control technologies with allied nations.

Global governance motives have been revealed in the United Nation's leadership for climate controls. The U.N.'s climate culture of eco-socialism has been exposed by members of its own Intergovernmental Panel on Climate Change (IPCC) established 1988. Several IPCC members have spoken in the pure Marxist-socialist principles of wealth re-distribution. A Chinese IPCC member stated that prosperous Western developed nation payments would be the key to success in global climate controls. And, a co-chairman of an IPCC working group has stated "One must say clearly that we redistribute *de facto* the world's wealth by climate policy.... One has to free oneself from the illusion that international climate policy is environmental policy."

Resigning his decades-long IPCC Chairmanship due to sexual harassment allegations in 2015, Dr. Rajendra Pachauri's resignation letter to the U.N. IPCC stated, "... sustainability of our ecosystems is more than my mission. It is my religion, my *dharma*." In Hinduism, dharma is the religious and moral law governing individual conduct. Sadly, the religious cultish and partisan politics of global warming have punished global prosperity and energy security without any practical benefit to humanity or wildlife. Pachauri was jointly awarded the *Nobel Peace Prize* with Al Gore in 2007. Like "green grifter" Al Gore, Pachauri has exploited his climate alarmism to acquire a global portfolio of business investments and relationships that has paid them millions of dollars from IPCC climate policies they enacted.

The IPCC is expected to shakedown developed countries for trillions of dollars in order to fund underdeveloped countries' climate controls in the guise of environmental and social justice. The "Paris Climate Accords" would extract over $100 trillion globally by the year 2100 to reduce global temperatures by only 0.08 degrees F (Bjorn Lomborg, 2017).

The Accords mandate that global climate temperatures must rise no more than 2 degrees Celsius (3.6 F) above preindustrial levels by the year 2100, and pursue a further limit of 1.5 degrees Celsius (2.7 F). It is clear that globalist, socialist ideology and cultish environmentalism have replaced prudent science and economics in climate policy.

Regarding air pollution control regulations, the EPA established "primary air pollutant" standards in the 1970s. These included carbon monoxide, lead, nitrogen oxides, ozone, particulate matter and sulfur oxides. These, as with most of the EPA pollution control regulations, are enacted to rest final enforcement responsibilities with the state governments. Therefore, each state has enacted and enforced its own air pollution programs consistent with EPA standard limits on air pollutant emissions. Climate concerns and government regulation of greenhouse gases did not exist until carbon dioxide (CO_2) was designated an "air pollutant" in 2007. Global environmental government air quality officials followed U.S. EPA's Supreme Court 2007 rulings that CO_2 is an air pollutant, and a hazard to human health and the natural environment. Note that this is the CO2 gas that you exhale in every breath, and that is essential for human and all animals' metabolism.

The 2007 Supreme Court ruling that CO2 is an "air pollutant" launched the global concerns to control the purported existential threat of "global warming" and "climate change."

Three decades ago, the United Nations Framework Convention on Climate Change (UNFCCC) was signed by President George H. W. Bush in Rio de Janeiro. The UNFCCC had one objective: to stabilize concentrations of greenhouse gases in the atmosphere to prevent "dangerous anthropogenic interference with the climate system." This objective incorporates three assumptions that collectively constitute a scientific and policy paradigm to control climate change.

1. Climate change is caused exclusively by human emissions of greenhouse gases, principally from the combustion of fossil fuels.
2. All of the climate impacts from burning fossil fuels are unambiguously bad for people and the planet.
3. The solution is the progressive, and preferably rapid, elimination of fossil fuels, requiring mankind to do without its main source of energy. (CFACT, August 2023)

Following the Kyoto Protocol 2012 expiration and in order to satisfy the 2015 Paris Accords target of global CO2 reductions, "Net Zero" carbon reduction mandates are causing global economic climate crises, particularly in the E.U. nation states known as "climate austerity." Net-Zero emissions, or "Net Zero," will be achieved when all greenhouse gas emissions released by human activities are counterbalanced by removing carbon from the atmosphere in a process known as carbon removal. Accordingly, carbon sequestration technology, where the atmospheric carbon dioxide greenhouse gas is captured in solution and stored underground. Sequestration has thus far proven infeasible. Various other novel CO2 air pollution capture technology experiments have been proposed.

Pres. Biden's "Green New Deal" has a lengthy, fanciful and costly wish list of policies to reduce climate change greenhouse gases.

Including:
Carbon sequestration,
Limit livestock flatulence,
Buy an electric car or scooter,
Convert public transits to electric,
Change natural gas appliances to electric,
Stop fossil fuel production in Alaska and Offshore,
Replace fossil fueled power grid with solar and wind power,
Up-zone residential areas with high-density multi-family housing,
Inflation Reduction Act provides $7,500 tax deduction for EV vehicle buyers,
Biden administration announces $7 billion in grants for plants to produce hydrogen fuel.

"The U.S. is supporting developing countries in taking stronger climate action –including providing $1 billion to the Green Climate Fund and requesting $500 million for the Amazon Fund and related activities – and invite other countries to join the U.S. and others in fully leveraging the multilateral development banks to better address global challenges, like climate change. President Biden will be joined by other leaders in new efforts aimed at accelerating progress in four key areas necessary for keeping a 1.5°C limit on warming within reach, specifically:

- Decarbonizing energy: Announcing steps to drive down emissions in the power and transportation sectors, including scaling up of clean energy, setting ambitious 2030 zero-emission vehicle goals, and decarbonizing international shipping.

- Ending deforestation of the Amazon and other critical forests: Working through the Forest and Climate Leaders' Partnership to mobilize public, private, and philanthropic support.
- Tackling potent, non-CO2 climate pollutants: Launching a Methane Finance Sprint to cut methane emissions and accelerating hydrofluorocarbon (HFC) phasedown under the Kigali Amendment.
- Advancing carbon management: Partnering with countries to accelerate carbon capture, removal, use, and storage technologies through a COP 28 Carbon Management Challenge to deal with emissions that can't otherwise be avoided.

(FACT SHEET: President Biden to Catalyze Global Climate Action through the Major Economies Forum on Energy and Climate, April 2023)

Biden's puppeteers want to spend another few million to buy votes and *green cred* with a cynical "American Climate Corps." The Corps' alarmists would recruit and exploit 20,000 proto-influencer Greta Thunbergs as climate-shamers to save us from the pseudo-existential threat of global warming (Atmospheric Thermodynamic Radiative Forcing). It is a fool's errand and a painful irony, that governments spend trillions to control global weather with weather-dependent wind and solar "renewable" energies. No science has concluded that human activities are the proximate and dominate cause of global climate change.

None of the much-heralded climate computer models account for the overwhelming, controlling influence of clouds (water vapor) over global temperatures and weather patterns that have evolved naturally over millions of years (Nobel Laureate Physicist John Clauser, Ph.D. July 2023). Clouds continually cover 45% to 95% of the earth. The feared carbon dioxide and methane greenhouse gases are mere parts-per-million influences. The only deaths attributable to climate hysteria will be those of world economies. The environmental movement has become the most densely-organized movement in human history. The over 15,000 environmental nonprofits will no doubt be further enriched by Climate Corps funding. When some one hears or reads the word "climate" as a threat... console yourself by just substituting the word "climate" with the word "weather.

Sustainable "renewable" energy sources have gained importance globally with mandates for CO2 reduction in the "War of Fossil Fuels." Designated renewable energy sources vary in the U.S. by state regulations. For example, California identifies wind, solar, small hydroelectric, geothermal and biomass as renewables. Note that nuclear power is omitted. Progressive greens have, in recent years, tried to close two major California coastal nuclear power plants. However, shortfalls in wind and solar replacement power sources have caused California to postpone or cancel the nuclear closures.

Progressive California government is the first state to propose that commercial businesses must publicly, annually report their greenhouse gas emissions. This, is a move to capture more emission fees associated with California's decades-old carbon-trading offset mitigation businesses, and to target industries with more climate-shaming propaganda.

Other climate-shaming regulatory schemes have arisen from the U.S. Federal Trade Commission (FTC) and global banking oversights. The FTC has regulatory "Green Guides" (16 CFR 260 et. seq.) that prohibit the advertisement of false or misleading "eco-claims." Manufacturer false claims of a product has "renewable," "biodegradable," "eco-friendly," "compostable," "sustainable," etc. performance can be an actionable FTC violation.

"ESG" (Environment, Social, Governance) has become an international business ratings scheme to publicly expose environmental, social and governance policies and performance of businesses for investment worthiness. The "environment" aspect would judge a business's compliances with climate political policies and *environmental justice.* The ESG scrutiny of corporate performance often exposes examples of *greenwashing,* where corporate funds are earmarked to promote conservation and environmental policies, but never actually meet conservation or environmental policy goals.

Exploiting ESG, *climate swaps* take existing emerging market countries' debt and either refinance it at a lower rate or extend the terms of the loan. The emerging nations then take the money raised and use it for biodiversity protection and climate adaptation. Climate swaps originated back in the 1980s, and have taken on new life as environmentalists encourage developing markets to spend more to conserve a counties lands and jungles. In developing countries, these swaps have enabled corruption and unrest, ESG audits and are often characterized by a lack of transparency. (New York Post, Oct. 2023)

CHAPTER 3

▼

GREEN ACTIVISM

Environmentalism, and its environmentalist believers activism, didn't become a potent public political movement until the 1960s and 70s in the U.S. when college campuses, brimming with idealistic baby boomers, were determined to make every new emotional twitch a political movement – a cause for revolution. This is when political movements, valid or not, became news media programming assets, and when anti-establishment and counter cultural influences became partisan media partners in a way that is largely taken for granted today. *Climate shamming* has become a common manipulation effort of green activism.

Environmental activism can have its own negative impacts; where the sanctimonious environmentalist's obsessive insistence that you change your lifestyle toward eco-purity begins to impair personal and professional relationships. Cultish, fetishistic devotions are displayed in radical environmentalism. The enduring negative impact of environmental activism is the constant politicization of your way of life, and limits to your freedom and prosperity. Ecopolitics are now consumed by chronic partisanship and scientific dysfunction.

Today, environmentalism mimics religion in its identification of humans as sinners against nature, its calls for personal redemption via expensive and disruptive lifestyle changes, its claims that environmental issues are moral imperatives, and, with the predicted global warming apocalypse, it has its biblically-proportioned Armageddon. As with most religions, environmentalism also has its false prophets and its extreme eco-terrorists.

Ecopolitics involves chronic partisanship and scientific corruption. Radical environmentalists have gone to the extremes of crashing public events and destroying public property. They've even defaced and damaged historic monuments and glued themselves to priceless works of historic museum art. "Green" acts of ecoterrorism have bombed medical research labs, burned industrial buildings and sabotaged highways, pipelines, land development projects and manufacturing. Many of these destructive acts are enabled by thousands of tax-exempt nonprofit (or NGO) environmental organizations of climate change "believers."

One large U.K. based radical eco-nonprofit, "Extinction Rebellion" (XR) claims, "We are in the midst of a climate and ecological breakdown. Now is not the time to ignore the issues; now is the time to act as if the truth is real. The science is clear. We are in the midst of a mass extinction of our own making and our governments are not doing enough to protect their citizens, our resources, our biodiversity, our planet, and our future." XR claims climate change activism in 88 countries. Some of these U.K. radical climate crusaders have pledged to forgo birthing babies due to their perceived existential threat of global warming.

Environmentalists claim moral authority by requiring your adherence to their cause of environmentalism. Your personal commitment and sacrifice for their fanciful and costly climate change imperatives provide your path to environmental enlightenment, eco-purity and ultimate green redemption. In open condescension, eco-warriors just want you to validate their "green self esteem" and "eco-piety" as heroic acts of citizenship.

Environmentalism has member believers whose daily lives involve a conscious self-reflection about human uses, conditions, consumptions, products, ambitions, ideology and survival as negative impacts upon natural systems. This self-reflective being, with rational to wildly irrational motivations, is an *environmentalist*.

Who are environmentalists? What do they believe? What do they want? Well, demographically, they're mostly middle to upper-middle class white, college-educated, socialistic, agnostic or un-religious, and seemingly well-intentioned people with lots of time on their hands. Most have never taken on the responsibilities of parenthood or running a business. They generally rely on academic and government institutions for their wisdom and values, and prefer group identity over self-reliance. Environmentalists have a utopian fixation on simplicity and even passivity as rules for living in a complex and energetic world. Ironically, simplicity and passivity make life poorer, not greener.

Radical environmentalists are intolerant of growth, prosperity and free enterprise. These eco-freaks rabidly resist measuring their goals against other critical concerns, like economics, national security, personal privacy and positive results. In their extremes, they constitute an "axis of antagonism" that provides no product or service in the furtherance of global prosperity. They reflexively detest competition, capitalism, political diversity, and prefer prescriptive globalism.

There are three basic and sustaining, yet fallacious, public perceptions that today's movement environmentalists (the *ecologista*) nurture and exploit. These public perceptions – indeed, misperceptions – persist because valid environmental analysis ultimately reduces to science, not feelings of compassion, guilt, fear, narcissism, political opportunism or fetishistic devotions. The public and its media have little interest in, or understanding of, science, and are easily manipulated by anyone with the facility for the hysterics that promote controversy and insecurity among the citizenry.

First, there is the flawed public perception that all local or regional environmental impacts, irrespective of their relative scale, endurance or ecology, are somehow cumulative to, and compounded with, all other known environmental impacts to result in a permanent catastrophic, irreversible global environmental collapse. There simply is no empirical example of any such terminal collapse occurring or likely to occur as a result of human activities. The expanding ripples on 19th Century writer/educator/preservationist Henry David Thoreau's *Walden Pond* were contained succinctly within the pond water, without impact to air or land. Even if one of the three natural resources of land, air or water is impacted, its condition does not necessarily add to, or compound with, other impacted resources in any grand global cataclysm. Provocative speculation that all individual impacts to the natural resources of land, air and water result in significant, irreparable ecological impairments is simplistically and functionally flawed.

Second, there is the public misperception that all living ecological systems are inherently fragile and mysteriously complex. Living ecological systems are powerful and wondrous forces whose resilience and adaptability far exceed man's relatively brief tenure and influence on Earth. Man has a tendency to conceitedly view all living systems in human consciousness – to anthropomorphize nature. Unlike man, nature acts only in the elegant efficiency of survival without ideology, morality, economics, politics, psychology, compassion or historical reflection. Nature is both intrinsically dominant and infallible – man is neither. Ecological systems have evolved 6xquisite assimilative capacities, mutations, bioremediations, dispersions, redundancies, regeneration, recovery and energy management capabilities. Ecological systems are not inherently fragile. The contrary is true. The scientific record of life on Earth demonstrates comprehensively and conclusively that ecological systems can and will sustain themselves infinitely in time, for and among all natural resources. This is true for atmospheric constituents associated with climate change.

Third, there is the public perception (again, misperception) that business and industry enterprises have a vested sinister interest and covert mission to destroy and pollute natural resources without regard for ecology or human health. Business and industry are motivated by profit, not pollution. Profit and pollution are ultimately incompatible goals for long term business and industry survival in 21st Century regulatory systems. Legitimate business and industry have adopted environmental compliance as part of their routine production and public relations objectives. Business and industry have demonstrated over the last 60 years that they will address environmental issues as long as regulatory controls are applied consistently among their competition, and as long as regulations are based in rational science with measurable positive economical outcomes. Long established and enforced regulations that limit industrial air pollutants coincidentally reduce greenhouse gases associated with global warming.

An original scholar in the theories of man's conflict with the sustainability of natural resources (the environment) was Thomas R. Malthus (1766-1834). Malthus was an English clergyman and economist. Malthus published an anonymous pamphlet in 1798 promoting the theory that human population increases geometrically, while food supplies can only increase arithmetically. Therefore, sooner or later, the growing gap between food supply and food demand must end in war, famine, and general human misery and extinction. Malthus simply argued that when mankind reaches the limits of nature – when it had farmed all the farmable land – mankind would starve. With the technology of 1798, Malthusian theory was both obvious and true in postulating that the ascent of man causes the collapse of everything else and that, in turn, destroys man too. Now, over 225 years later, clearly Malthus was in error because he grossly underestimated man's ever-evolving ingenuity. Human resourcefulness was left out of his equation.

Born and raised on a farm in Springdale, Pennsylvania, Rachel Carson wrote a seminal book on the overuse of U.S. pesticides in agriculture titled *Silent Spring* in 1962. DDT and its derivative chlorinated hydrocarbons were agricultural pesticides developed in the late 1930s. DDT was widely used in farming to improve crop yields, and was found to control almost any type of insect pest. These chlorinated hydrocarbons are quite resilient in the environment, having decomposition half-lifes of 10 to 15 years. Though not an ecologist, Carson's book implicates reckless worldwide pesticide use in the unraveling of ecosystem food chains in very personal, almost romantic language. Carson predicted that pesticide use would eventually contaminate world drinking water supplies. DDT use in the US was banned by Federal Law in 1972. Carson's prediction, fortunately, was wrong. (Green Gone Wrong, Paul Taylor 2001)

A more consequential environmental scholar, probably because he wrote during the 1960s and 70s era when college campus political movements *flowered*, is Stanford University biologist Paul Ehrlich. In 1968, Ehrlich wrote a bestseller titled *The Population Bomb*. Ehrlich's book embraces and extends the Malthusian theory toward the irrational. Ehrlich's 1968 doomsday vision stated, "The battle to feed all of humanity is over. In the 1970s and 1980s, hundreds of millions of people will starve to death in spite of any crash programs embarked upon now." Time has proven Ehrlich to be flamboyantly and profoundly wrong. He admonished that nature will take its revenge against mankind's abuses. Further, Ehrlich attributes the AIDS epidemic to the "deterioration of the epidemiological environment which is quite directly related to population size as well as to poverty and environmental deterioration." Ehrlich appears to have stumbled from his basic life science expertise into elitist psycho-socio-babble, to his professional discredit.

U.S. Senator, Vice President and presidential aspirant Al Gore authored a 400-page tome on everything environmental titled *Earth in the Balance: Ecology and the Human Spirit*, that became a "national bestseller." Gore, in writing the book, said he was willing to risk his entire political career on the issue of *the environment*. His original 1990 title for the book was "World War III" – ironic for a 2007 recipient of the *Nobel Peace Prize*. Gore's emphasis was that to attain his version of global environmental rectitude would require the commitment and sacrifice of world war. *Earth in the Balance* rambles maniacally among premises of over population, nature's spirit, technophobia, consumptionism, species extinction, and introduced us to the term *global warming* in an attempt to circumscribe an "environmental holocaust without precedent," and to position Gore as the political leader whose insight will save the planet.

Gore exposes himself as either profoundly confused, or cynically manipulative, about the meaning of "technology." He writes "[G] overnment, as a tool used to achieve social and political organization, may be considered a technology, and in that sense self-government is one of the most sophisticated technologies ever created." His further abstractions equate technology with "spoken language," and even "the human body." Gore also elaborates on how technology is not necessarily science. One should always be suspicious when politicians begin to bend the meaning of words. For clarity, please observe that Webster's Dictionary defines technology as "applied science." Gore also calls for science and religion to be "reunited in the service of the environment."

Gore, in keeping with the world war analogy of his environmental crusade, promotes vast government programs such as a "Strategic

Environmental Initiative" and "Global Marshall Plan." His Global Marshall Plan would act to stabilize world population, develop environmentally appropriate technology, measure environmental impacts in economic terms, develop international environmental regulatory programs and develop a global environmental education program. His Strategic Environmental Initiative was named to imply an environmental equivalent of the "Strategic Defense Initiative" (SDI), the crash program to develop a series of technological breakthroughs focusing on a common military objective, which Gore opposed as senator. Gore's Strategic Environmental Initiative would be a global "program that would discourage and phase out older, inappropriate technologies and at the same time develop and disseminate a new generation of sophisticated and environmentally benign substitutes." This, from a politician who clearly does not understand the definition of the word "technology."

Gore further writes to introduce the divisive environmental concepts of "environmental justice" and "sustainable development" in pandering to the *ecologista*. These latent environmentalist concepts are attempts to leverage social issues of immigration, social justice, class warfare, racism and big business bashing for pure political patronage.

In 2007, Al Gore received an *Oscar* for the movie documentary *An Inconvenient Truth* that was a cinematic version of his slideshow tour, wherein he conceitedly claims that "... the debate about global warming is over." Gore's "truths" have been widely discredited by climate scientists. Gore, far from ending the debate, actually started the debate about global warming. *Vanity Fair,* magazine of social record, validation and wisdom for our plutocrats, had a "Green Issue" that attempted to rally eco-consciousness with the fanciful cover headline – "A Threat Greater Than Terrorism: Global Warming." Ironically, this headline is both a denial of the horrors of *9/11*, and an admission that the U.S. is winning the war on terror. The magazine presents Al Gore, among other *green elites*, as the moral arbiter and oracle on all things environmental – including global warming.

A more recent example of the endless desire to identify with *things environmental* is the pseudo-scientific, psycho-social, co-optation called *ecopsychology*. Every political movement has its psychological dimension. This is no less true for the environmental movement. Persuading people to alter their behavior always involves probing personal motivations and values. Political activism begins with asking what makes people tick. What does the public want, fear and care about? How do we get and hold the public interest? How much can people take; what are their priorities? Have activists overloaded the populace with anxiety and guilt? Ecopsychology endeavors to address the problem of effective communication with the general public in order to meet the demands of the "environmental revolution." Ecopsychology claims to "redefine sanity within an environmental context," to re-examine the human psyche as an integral part of "the web of nature" or ecosystem. Ecopsychology presumes to "bring together the sensitivity of psycho-therapists, the expertise of ecologists and the ethical energy of environmental activists." Here is a perfect example of the abstraction of a physical science discipline (i.e., environmental science) with a soft pseudo- science discipline (i.e., psychology, sociology, etc.). This is sophistry, not science, and certainly no basis for making prudent management decisions about our natural resources or climate change.

The fuzzy regulatory area of non-, or not-for-, profit tax-exempt organizations (or NGOs) deserves scrutiny as regards the global environmentalism and its groups. Most of the well-known environmental groups operate as tax-exempt organizations. This tax-exempt status means that nonprofits are effectively subsidized by all other taxpayers. Purportedly, tax-exempt status infers that these groups are taxpayer subsidized because they act in the *public interest* to enhance and protect *public welfare*, and they therefore deserve the *public trust*.

According to the U.S. Internal Revenue Service (IRS), general categories of tax-exempt organizations include those organized and operated exclusively for one or more of the following purposes: religious, charitable, scientific, testing for public safety, literary, educational, protection of children or animals, or amateur sports promotion. Few of the dominant tax-exempt environmental organizations can be seen to strictly qualify under one or more of the foregoing IRS categories. Many of the prominent eco-nonprofit organizations frequently jeopardize their tax-exempt status under the IRS "propaganda" prohibitions. The IRS Code states, "… (organization is exempted provided) no substantial part of the (organization's) activities… is carrying on propaganda or otherwise attempting to influence legislation or intervene in any political campaign on behalf of any candidate for public office." Propaganda is defined as ideas, facts or allegations spread deliberately to further a cause or to damage an opposing cause. Eco-nonprofits have been cautioned by the IRS for lobbying and propaganda activities that exceed the "no substantial part of activities" tax exempt status standard under the IRS Code.

Recently, the IRS began increased oversight of the approximately 1.3 million nonprofit organizations operating in the U.S. under IRS Code Section 501(c)(3). Thousands of charitable organizations routinely endorse political candidates in local, state and federal elections. These endorsements come mainly from environmental, civil rights, labor and other "research and education" nonprofits. Nonprofits have grown out of control and unaudited over the last thirty years, coincident with political polarity in America – environmental nonprofit groups grew from 2000 to 4000 during the 1990s.

Eco-groups take tax-deductible donations from you, foundations and corporations to operate global fear-mongering campaigns with theoretical problems of pollution, species extinction, climate change, etc. These eco-groups also lobby intensely for costly government regulatory controls that are particularly onerous for those least-privileged persons and families among us. This has become a form of "green tyranny," complicit with the *enviro-industrial complex*; and, leading ultimately to *climate austerity*.

Tax-exempt nonprofits include the institutions of foundations, churches, the arts, social/animal welfare, civic service clubs, sports, education, scientific, unions, financial retirement services and ubiquitous environmental causes. Such organizations fall under the Federal IRS Code Section 501(c) in twenty-four subsection classifications. These organizations pay no income taxes, and have various, but few, political lobbying restrictions. The late political columnist and psychiatrist Charles Krauthammer has said, "American nonprofits are out of control quasi-governmental organizations." Taxpayer-subsidized "nonprofits" are slush funds for political mischief and institutional fraud. Statistics from the Nonprofit Times state that the total assets of U.S. nonprofit organizations are approximately $6 trillion, greater than the GDP of Japan. 80% of nonprofits' revenue comes from governments. There are approx. 1.5 million U.S. nonprofits, 111,000 in California alone. Nonprofits include 1.4 million "charitable nonprofits," and approximately 130,000 foundations. The number of nonprofits increased by about 75% between 2000 and 2016, and environmental nonprofits increased from 2,000 to 4,000 during the Clinton administration alone. More than 12 million people are employed in nonprofits, more than 10% of the U.S. workforce. Nonprofits spend about $2 trillion annually, including approx. $9 billion on salaries and benefits. 92% of nonprofits have annual budgets of less than $1 million. (The Nonprofit Times, 2022)

The sound and meaning of the term "nonprofit" are antithetical to "capitalism." 501 (c) tax-exempt nonprofit laws were first enacted in the 1954 under the U.S. Revenue Act. Nonprofits' proliferation and performance as quasi-government entities are largely left to state law enforcement. Nonprofit incomes are not taxed, and they cannot declare common bankruptcy, but must declare dissolution or transfer without any asset sale proceeds or capital gains. Essentially, nonprofits cannot be bought or sold as commercial transactions. Appling today's corporate income tax rate to the tax-exempt $6 trillion nonprofit assets could add over $1 trillion in tax payments to the U.S. Treasury.

Nonprofit 501(c) regulations should be updated and consolidated to control lobbying and political corruption, qualifying-purpose performance failures and illicit tax evasion. Taxpayers are paying twice to subsidize tax-exempt nonprofits - once by the government tax funds donated to nonprofits (80%), and again by your taxes paid to the U.S. Treasury to replace income taxes not paid by tax-exempt nonprofits. Progressive-government administrations have become reliant upon the nonprofits as partisan political lobbying operations.

Some environmentalists promote the anti-capitalist doomsday threats of a movement that now exists largely as a partisan political power constituency. Green nonprofits are the source of numerous lawsuits and Democrat funded political campaigns involving environmental issues.

Most litigious U.S. environmental non-profit organizations (IRS Sec. 501, C, 3 Form 990s)

- Center for Biological Diversity – Exec. Comp. over $200,000. Annually
- Environmental Defense Fund – Exec. Comp. over $600,000. Annually
- Greenpeace Inc./Fund – Exec. Comp. over $200,000. Annually
- National Resources Defense Council – Exec. Comp. over $500,000. Annually
- Sierra Club (Earth Justice) – Exec. Comp. over $300,000. Annually

Today, eco-groups punish prosperity by exploiting government regulations, unsettled science, gullible media and your *green guilt*. Elite green groups are *nonprofits* that profit as lawyers writing and lobbying costly eco-laws, awarding lavish grants to green-biased researchers, and suing government and industry for excessive regulations and monetary judgments with *citizen* standing in federal courts. Environmentalists pose in virtuous green rhetoric, but are never held accountable for their damage if measured against the critical criteria of economics, national security and scientific rigor. Eco-groups must now take responsibility for the destructive social and economic impacts of their myopic and ideologically-driven causes.

10 worst eco-group causes ...

1. **Against** nuclear power plants that provide reliable, renewable electric power without greenhouse gases, and that can replace coal-powered plants while reducing US dependency on foreign oil;

2. **Against** safe, new domestic oil and natural gas exploration and production on our public lands, offshore and in the Alaskan tundra where US petroleum reserves are potentially the third largest in the world;

3. **Against** new petroleum refineries in the US where refined fuels production capacities are little changed in 30 years;

4. **Against** public forest land management and roads that would allow containment of destructive and polluting wildfires;

5. **Against** new border security control facilities that would control illegal immigration, and human and drug trafficking that cost California and other border states billions of dollars per year in crime and welfare expenditures;

6. **Against** new public infrastructure such as roads, bridges, pipelines, transport hubs, dams, power plants and power

transmission facilities that promote human growth, safety and economic vitality;

7. **Against** research and commercialization of genetically-enhanced crop seeds that require less polluting fertilizer, less water, less land and less time to grow the food grains essential to the human diet worldwide;

8. **For** stricter car mileage regulations that will inflate US vehicle prices and make our struggling carmakers less competitive in foreign markets;

9. **For** a national *cap-and-trade* carbon-taxing system that has failed in Europe, and that will add broad and uncertain costs to all products and services in a time of global economic recession;

10. **For** *renewable energy* such as wind power, solar power, geothermal, and hydrogen and biomass fuels, where the cost of wind and solar electric power is five times that of conventional coal-fired electric power.

Under the Equal Access to Justice Act where often the green litigants prevail, U.S. government institutions (and ultimately the U.S. taxpayer) pay the legal fees of these eco-groups. While receiving $20 million in grants during the last decade, the Environmental Law Institute publishes a "Citizen's Guide" to promote these parasitic *citizen suits* against EPA, the Dept. of the Interior, the Dept. of Energy and others. A dozen eco-groups are responsible for over 4,000 *citizen suits* filed against EPA and the Dept. of the Interior in the last decade.

Have no doubt that environmental awareness has benefitted both the human and natural environment with diligent science-based regulatory controls. And, we all pay more for every product, service, food, energy and activity with each new, restrictive and prescriptive environmental regulation.

CHAPTER 4

▼

ECOPROPAGANDA

Politics and truth seldom occupy the same space anymore. When the environmental movement found a political base, it began to leave the truths of scientific rigor in natural resource management policy behind as too slow, too cumbersome. In the climate of ecopolitics, the environmental movement has lost its way and legitimacy in the 21st Century. Environmentalism has become just another partisan means to a political end.

Over the last 60 years, ecopropagandists have done a good job of frightening and shaking the general populace into first the hysteria of "the sky is falling," then more recently into resentment over the exaggerated daily claims of environmental apocalypse, until today where warning-battered rational people are barely listening to environmentalists. After years of being bombarded by ever more dire ecological prophecies, of which none has materialized, citizens have grown more and more skeptical of environmentalist predictions and protests. The alarmist theories that environmentalist made fashionable during debates over acid rain, toxic groundwater, nuclear winter, over-population and species extinction, have backfired. Citizens have now adopted a "selective deafness" as a first line of defense against wild claims of environmental disaster.

Ecopsychologist Theodore Roszak noted in his 1993 book, *The Voice of the Earth*, that the environmental movement might have overutilized shame-and-blame tactics in its approach to the public. And, that the public may become particularly vulnerable to right-wing conservative attempts to instigate a "green backlash." Roszak theorizes that the green backlash may provide people an opportunity to avoid feelings of guilt and helplessness, and to attack environmentalists who make them feel that way. Roszak's bizarre musings reveal environmentalisms perverse and cynical manipulative devices of promoting public guilt and helplessness (i.e., *victimization*).

As the focus of environmentalists moves from once-immediate dangers now under control, to more abstract matters of aesthetics or sustainability or global warming, latent economic class conflicts are beginning to erupt. Pollution controls often impose highly regressive costs according to socio-economic class. For example, in the early 1990s, the cost burdens of Southern California's aggressive air quality management plans were estimated to have a three times greater impact upon the region's poorest households than on the wealthiest. Environmentalists dismiss such economic irritants by arguing that a better environment helps everyone. The proverbial ecological crisis notwithstanding, the adverse health consequences of reduced economic opportunities for the poor vastly overwhelm any environmental benefits they may enjoy from, say, marginally cleaner air quality. U.S. air pollution reduces average life expectancy by approximately 30 days. Poverty strips away 10 years in life expectancy.

The experience of 40 years of environmental controls in the U.S. is testament to conserved natural resources and solved pollution problems, and conclusively demonstrates that growth-oriented economies (i.e., free-market democracies) actually do a better job of managing natural resources than a society run on the myopic principles and utopian-directed theories of environmentalist dogma.

Environmentalists are blindly resentful of the broad growth in economic opportunities that come from free enterprise. Environmentalists see themselves as heroic (if not messianic) figures in a movement that is more socialistic than problem solving. This is where the environmental ideology can take on the trappings of a religious crusade. The movement transcends the need for truth to aid proselytization.

Global warming has spawned fanatic believers in abstract theories, mythologies and mysticism in today's environmental movement that reach a near fetishistic fervor. Sadly, the productive scientific motives and technologies that solve environmental problems have been bypassed and replaced by perverse and partisan political dogma, demagoguery and fearmongering propaganda. The *green* movement began in prospering 20th-Century democracies such as the U.S. and Western Europe; because their affluence absorbed the costs of environmental controls. The environmental movement grew to become the largest, most densely organized public cause in human history. Today the movement increasingly mimics religion in its identification of humans as *sinners* against nature, its calls for personal redemption via lifestyle changes, its claims that environmental issues are now moral issues, and, with the predicted global warming apocalypse, it has its biblically-proportioned *Armageddon*. As with other religions, environmentalism also has its *false prophets,* and its extremist followers in the form of animal rights cults and *eco-terrorist* activists.

The religion of environmentalism has a simple, godless good-versus-evil orthodoxy: nature does good, man does bad. This basic belief enables anti-capitalists and global-socialist political operatives. The faithful congregations of environmentalism are found under the roofs of the United Nations, the European Union, the global Green Party, the Democrat Party in America, and the over 15,000 nonprofit eco-groups. These organizations are the dogmatists for the eco-religious – environmentalism is their *de facto established religion*. Further, these are the institutions from which environmentalism's dogma and false prophets exploit political opportunity. Eco-activist's claim moral authority in requiring your adherence to the cause of environmentalism. Your personal commitment and sacrifice for their fanciful moral imperatives such as *global sustainability, smart* _____ (you fill in the blank), *carbon footprinting* and *environmental justice* provide your path to environmental enlightenment, purity and ultimate redemption. Many of these feel-good goals have become common policy language in local, state. national and international environmental regulations. These policy goals are fed wholesale to our children in *politically-correct* public school curricula where the mention of traditional religions has been banned for decades – kids are programmed to *green worship*. Global news media have been a willing and gullible accomplice in the hysterics of environmentalism, and in particular, regarding global warming.

As with all religions, environmentalism has the harmful impacts of false prophets, cult extremists, and conflicts with proven science. No where in the history of the environmental movement have the false prophets and extremists been more duplicitous than in the eco-propaganda and political exploitation of the theory of global warming. Wild doomsday speculation from green group fund raising propaganda have distorted the issues of climate change far beyond rational scientific discovery or discourse. Recent published science concludes that the Earth is warming, but that the crushing global economic impact of fully-implemented, for example, "Net Zero" would only reduce global warming gases by less than one percent during the 21st Century. Contrary to religion, science is not about belief, hyped hypotheses, corporate conspiracies or political opportunism. The applied sciences that enable and protect your everyday activities are about repeatable, measurable proof of a theory for cause and effect concerning physical phenomenon using the *scientific method*. Policymakers must resist environmentalism's seduction of short-term political gains, and await the conclusive science that properly analyses global climate change. Short-term leverage of environmentalism for political expediency will only hasten and lengthen a global economic collapse, with far greater human suffering, conflict and pollution than any of the worst-case scenarios being proliferated by environmentalism's 21st-Century climate crusade.

In order to glean some truth from any activist propaganda, one must start from the cautious position that the first casualty of activism is the truth – truth is secondary. Activist, including environmentalists, rhetoric is tactically reliant upon exaggerated dangers and inflammatory word use to erect counterfeit arguments and promote bumper sticker platitudes as moral authority for social engineering. Typically, the *zero sum* false dilemma proposed is that of "it's only man or nature" that will survive the current environmental disaster. (The Guilder Technology Report)

Anyone with only a rudimentary knowledge of the living systems of Planet Earth, that include both man and nature, understands that nature, with or without man, shall ultimately prevail just as it has for over 4 billion years prior to man's unceremonious and primitive appearance on Earth about 3 million years ago.

The founding beliefs of environmentalists are that of scarcity, of limits to growth and therefore, "the sky is falling," "the end is near," "catastrophe is just around the corner" fear-mongering proclamations. Their basic and transparently political theory for the collapse of the Earth's ecosystem is blamed mainly on the American capitalist way of life. One can only surmise that the operative environmentalist theory is that the high quality of western capitalistic life must somehow be a threat to the Planet Earth.

Where is the truth? Recent scholars in the true environmental sciences have coined the term *trans-science* to describe the study of phenomena that are too large, too diffuse, too rare, too distant or too long term to be resolved by reliable scientific methods. This trans-science defines the current boundary of understanding climate science and global warming. Exploiting trans-science, environmentalists release so-called "studies" that bypass the scientific journals and peer review, and go straight to sympathetic, issue-hungry and largely gullible journalists. The due diligence of true scientific cause-and-effect findings is neglected. (Hudson Institute April 1999)

Of course, without each of us applying some critical examination and context to these waves of scary news, the complex, remote and invisible threat will make you uneasy as it is designed to do. It is also designed to create civic anxiety, demonize business and industry, and thus promote a dependency on government to solve a problem that may not exist in the priorities of our lives or global sustainability. Consensus in legitimate science has always led to rational truth about physical phenomena such as climate change. Today, the radical environmental movement is more alchemy than altruism. Today, the media is more concerned with the controversy that attaches to environmentalism, rather than truth telling about the state of the environment.

American courtrooms and government regulatory agencies have been overwhelmed by health scares linked to environmental issues – pesticides, ozone depletion, electromagnetic waves from cell phones, etc. Too often, government environmental researchers operate as a self-selected, insular, academic group with an inherent bias toward identifying environmental risks. Often, this leads them to become personally, emotionally and financially wedded to their own theories, and scientific objectivity is lost along with the truth. (American Council on Science and Health, July 1999)

Today, a successful government regulatory scientist may find possible environmental problems, and then publicize them as probable environmental crises to get the attention of legislators for follow-up work and renewed funding.

Often the incentives in government are to save, rather than solve the problem, and thereby save bureaucratic power and it's vast taxpayer-supported *civil service* employment opportunities and its massive partisan voting blocks. (The Wall Street Journal, Aug. 1999)

Today, there is an unholy alliance of environmentalists, media and regulatory bureaucrats worldwide that conjures up environmental evils, and dresses them up as science in a system that is perfectly evolved to fund and grow the global government establishment. There are legions of academics and regulatory scientists whose occupation it is to invade, critique, punish, and ultimately dictate your lifestyle – ecopolitics reigns.

Green marketers have developed sophisticated schemes to sell an avalanche of eco-friendly and sustainable products. They believe that consumers will support green products, when given the right information. The marketing themes, as with for other products, must emphasize an immediate, practical and emotionally-compelling benefit in a way that relates to the daily lives of the target buyer. Green marketers use detailed research survey data in an attempt to develop hard-hitting messages that make environmental protection tangible and relevant. They also deploy these messages through highly-leveraged partnerships with other products, institutions and media that are already a part of the consumer's lives. (Eco America Website, Jan. 2008)

The insurance industry's clout is sizable. It's the second-largest industry in the world in terms of assets, and has a direct link to most homeowners and businesses. It insures coal-fired power plants as well as wind farms, so it can influence the power industry's cost structure. With its financial muscle, the industry can require the use of new financial instruments designed for companies to trade greenhouse-gas emissions in the same way that commodities are bought and sold.

The insurance industry has the ability to change behavior, costs and policies, and to squeeze millions of clients. Some consumers are already noticing a negative effect in their insurance. While the insurance giants are echoing *green alarmists*, they are also attempting to cut coverages and raise premiums on individuals and business by advanced speculation on the uncertain and incomplete scientific cause and effect of global warming.

Psychiatrists, psychologists and sociologists have long postulated that symptoms of irrational fear and anxiety increase when political and economic systems are most unstable and unprosperous. This is why institutionalized enviro-mythmaking and climate-shaming have proved to be quite powerful, persuasive and pernicious. The involved institutions are mass media news sources such as TV, radio, the internet, magazines and newspapers. The basic, raw material for mass media is controversy, to feed the activity of environmental fear mongering. University of Southern California Professor of Sociology, Barry Glassner, wrote a seminal and best-selling book on fear mongering in 1999. (The Culture of Fear, Glassner 1999)

Glassner debunks (among others) the 1990s enviro-myth of the dangers of U.S. schools containing asbestos. Government regulators estimated that one-third of the nation's schools contained asbestos insulation that when inhaled over long periods of time can cause lung cancer. That school kids would be exposed to the asbestos health risk, became a public outrage. U.S. schools spent over $10 billion to remove school asbestos even though its removal posed a greater health hazard risk than allowing the asbestos to remain installed and immobile. In this case, media engaged relentlessly in the school asbestos health scare as fear mongering; as they have with other health scares such as AIDS, Dow Corning silicon breast implants, Gulf War Syndrome, road rage and now global warming. Why do fear mongering media campaigns take hold? Why do media and their audiences get drawn to one hazard after another?

Fear mongering motivates people to 1) correct a moral offense or 2) to criticize a disliked group or institution. (Illness as Metaphor, Farrar, Straus & Giroux 1989)

Health hazards, at any degree of injury or prevalence in the population, are deemed to be morally unacceptable, whether merely perceived or real in scientific medical terms. Witness the school asbestos example above as a motivating environmental health hazard. Some of this motivation arises out of a collective cultural narcissism or sense of entitlement; because real or not, health risks are innately personal and have resonance with our basic survival instincts.

The second fear mongering motivation to criticize or discredit a disliked group or institution came into full influence in western culture during the 1960s and 70s. Then until now, demonizing groups or institutions has become the activists' sport and even occupation to leverage environmentalism's distrust of corporate free enterprise. Given the high threshold for motivating moral outrage and the seeking of personal redemption via public political protest that have characterized environmentalism, they are not likely to be motivated by an aforementioned moral offense in response to environmental issues. Rather, the large, faceless target is big business. Some of this antipathy has its roots in the historic labor union conflicts with big business and pure Marxist socialism.

The isolated, dramatic, personal anecdote of some environmental issue is the *smoke* that is fanned into the *flames* of public outrage by media for government to regulate big business, be it industrial manufacturing, mining, oil, home builders or their financiers. With the flames lit, the more public talk there is about the reported environmental issue, the more likely are other accident-monitoring agencies such as police and insurance to collect similar examples of the reported issues that they would have ignored altogether or classified differently prior to the reports. Psychologists call this the "Pygmalion effect." Further, fear mongering relies upon what psychologists refer to as the *availability heuristic*, where people judge the significance of an isolated issue by how readily it comes to mind. When we are presented with a survey that polls the relative significance of an issue, we are likely to give greatest significance to whatever the media emphasizes at the moment because that issue tends to come to mind. (The Culture of Fear, Glassner 1999)

Fear mongers make their scares all the more credible by having professional spokespersons or *victim-cum-experts* to spread pseudo-scientific or dramatic testimonial information about an environmental issue. Professional narrators also play an important role in transforming implausible environmental threats into the "disaster de jour." The rantings of alarmist newscasters and the glorification of wannabe experts are two tricks that expose the fear mongers manipulative ploy. In addition, the use of tragic anecdotes in place of statistically-tested scientific evidence, and the conflation of isolated incidents into trends or conspiracies are an attempt to exploit deeper cultural anxieties, divides and hatreds, and even personal paranoia. The simple answer as to why we are exposed to so many persistent and irrational fears is that immense political power and financial rewards await those who tap into our moral insecurities and supply us with symbolic substitutes. (The Culture of Fear, Glassner 1999)

The U.S. wastes tens of billions of dollars and human resources every year on fear monger-promoted enviro-mythological issues, including research and technology, and on victim compensations for *metaphorical illnesses*. Metaphorical illnesses include Gulf War Syndrome, multiple chemical sensitivity and silicon breast implant disorders where people justify their personal fears, prejudices, hardships and political ideologies in the absence of determinative scientific explanations by projecting themselves into a public class of victims and litigants. This today's *victim as virtue* media epidemic. (Illness as Metaphor, Farrar, Straus & Giroux 1989)

Cause and effect findings of environmental impacts can be reliably established when the significance of valid scientific data are subjected to statistical analysis – it tells you whether sufficient quality and quantity of data are available to determine the cause and effect theory tested. Scientific cause and effect cannot be determined by coincidence, personal anecdotes, circumstantial evidence or preponderance of court evidence which are prevalent legal standards, not scientific conclusions. Scientific cause and effect has not yet been developed for accurate projection or remediation of the possible impacts of global warming.

The following over 600 propagandist items have been published as being caused by global warming:

Acne, agricultural land increase, Africa devastated, African aid threatened, Africa hit hardest, air pressure changes, Alaska reshaped, allergies increase, Alps melting, Amazon a desert, American dream end, amphibians breeding earlier (or not), ancient forests dramatically changed, animals head for the hills, Antarctic grass flourishes, Antarctic ice grows, Antarctic ice shrinks, anxiety, algal blooms, archaeological sites threatened, Arctic bogs melt, Arctic in bloom, Arctic ice free, Arctic lakes disappear, asthma, Atlantic less salty, Atlantic more salty, atmospheric defiance, atmospheric circulation modified, attack of the killer jellyfish, avalanches reduced, avalanches increased, Baghdad snow, bananas destroyed, bananas grow, beetle infestation, bet for $10,000, better beer, big melt faster, billion dollar research projects, billions face risk, billions of deaths, bird distributions change, bird visitors drop, birds return early, birds driven north, blackbirds stop singing, blizzards, blue mussels return, bluetongue, boredom, bridge collapse (Minneapolis), Britain Siberian, British gardens change, brothels struggle, bubonic plague, budget increases, Buddhist temple threatened, building collapse, building season extension, bushfires, business opportunities, business risks, butterflies move north, camel deaths, cancer deaths in England, cardiac arrest, caterpillar biomass shift, cave paintings threatened, challenges and opportunities, childhood insomnia, Cholera, circumcision in decline, cirrus disappearance, civil unrest, cloud increase, cloud stripping, cockroach migration, cod go south, cold climate creatures survive, cold spells (Australia), computer models, conferences, coral bleaching, coral reefs dying, coral reefs grow, coral reefs shrink, cold spells, cost of trillions, cougar attacks, cremation to end, crime increase, crocodile sex, crumbling roads, buildings and sewage systems, cyclones (Australia), damages equivalent to $200 billion, Darfur, Dartford Warbler plague, death rate increase (U.S.), Dengue haemorrhagic fever, dermatitis, desert advance, desert life threatened, desert retreat, destruction of the environment, diarrhoea, disappearance of coastal cities, diseases move north, Dolomites collapse, drought, drowning people, ducks and geese decline, dust bowl in the corn belt, early marriages, early spring, earlier pollen season, Earth biodiversity crisis, Earth dying, Earth even hotter, Earth light dimming, Earth lopsided, Earth melting, Earth morbid fever, Earth on fast track, Earth past point of no return, Earth slowing down, Earth spinning out of control, Earth spins faster, Earth to explode, Earth upside down, Earth wobbling, earthquakes, El Niño intensification, end of the world as we know it, erosion, emerging infections, encephalitis, equality threatened, Europe simultaneously baking and freezing, evolution accelerating, expansion of university climate groups, extinctions (human, civilisation, logic, Inuit, smallest butterfly, cod, ladybirds, bats, pandas, pikas, polar bears, pigmy possums, gorillas, koalas, walrus, whales, frogs, toads, turtles, orang-utan, elephants, tigers, plants, salmon, trout, wild flowers, woodlice, penguins, a million species, half of all animal and plant species, mountain species, not polar bears, barrier reef, leaches), experts muzzled, extreme changes to California, fading fall foliage, famine, farmers go under, fashion disaster, fever, figurehead sacked, fir cone bonanza, fish catches drop, fish downsize, fish catches rise, fish stocks at risk, fish stocks decline, five million illnesses, flesh eating disease, flood patterns change, floods, floods of beaches and cities, Florida economic decline, flowers in peril, food poisoning, food prices rise, food prices soar, food security threat (SA), footpath erosion, forest decline, forest expansion, frostbite, frosts, fungi fruitful, fungi invasion, games change, Garden of Eden wilts, genetic diversity decline, gene pools slashed, giant squid migrate, gingerbread houses collapse, glacial earthquakes, glacial retreat, glacial growth, glacier wrapped, global cooling, global dimming, glowing clouds, god melts, golf Masters wrecked, Gore omnipresence, grandstanding, grasslands wetter, Great Barrier Reef 95% dead, Great Lakes drop, greening of the North, Grey whales lose weight, Gulf Stream failure, habitat loss, Hantavirus pulmonary syndrome, harvest increase, harvest shrinkage, hay fever epidemic, hazardous waste sites breached, health of children harmed, heart disease, heart attacks and strokes (Australia), heat waves, hibernation ends too soon, hibernation ends too late, homeless 50 million, hornets, high court debates, human development faces unprecedented reversal, human fertility reduced, human health improvement, human health risk, human race oblivion, hurricanes, hurricane reduction, hydropower problems, hyperthermia deaths, ice sheet growth, ice sheet shrinkage, illness and death, inclement weather, infrastructure failure (Canada), Inuit displacement, Inuit poisoned, Inuit suing, industry threatened, infectious diseases, inflation in China, insurance premium rises, invasion of cats, invasion of herons, invasion of midges, island disappears, islands sinking, itchier poison ivy, jellyfish explosion, Kew Gardens taxed, kitten boom, krill decline, lake and stream productivity decline, lake shrinking and growing, landslides, landslides of ice at 140 mph, lawsuits increase, lawsuit successful, lawyers' income increased (surprise surprise!), lightning related insurance claims, little response in the atmosphere, lush growth in rain forests, Lyme disease, Malaria, malnutrition, mammoth dung melt, Maple syrup shortage, marine diseases, marine food chain decimated, marine dead zone, Meaching (end of the world), megacryometeors, Melanoma, methane emissions from plants, methane burps, melting permafrost, Middle Kingdom convulses, migration, migration difficult (birds), microbes to decompose soil carbon more rapidly, monkeys on the move, Mont Blanc grows, monuments imperiled, more bad air days, more research needed, mortality increased, mountain (Everest) shrinking, mountains break up, mountains taller, mortality lower, mudslides, National security implications, natural disasters quadruple, new islands, next ice age, Nile delta damaged, noctilucent clouds, no effect in India, Northwest Passage opened, nuclear plants bloom, oaks dying, oaks move north, oblivion, ocean acidification, ocean waves speed up, opera house to be destroyed, outdoor hockey threatened, oyster diseases, ozone loss, ozone repair slowed, ozone rise, Pacific dead zone, personal carbon rationing, pest outbreaks, pests increase, phenology shifts, plankton blooms, plankton destabilised, plankton loss, plant viruses, plants march north, polar bears aggressive, polar bears cannibalistic, polar bears drowning, polar bears starve, polar tours scrapped, porpoise astray, profits collapse, psychosocial disturbances, puffin decline, railroad tracks deformed, rainfall increase, rainfall reduction, rape wave, refugees, reindeer larger, release of ancient frozen viruses, resorts disappear, rice threatened, rice yields crash, riches, rift on Capitol Hill, rioting and nuclear war, rivers dry up, river flow impacted, rivers raised, roads wear out, rockfalls, rocky peaks crack apart, roof of the world a desert, rooftop bars, Ross river disease, ruins ruined, salinity reduction, salinity increase, Salmonella, salmon stronger, satellites accelerate, school closures, sea level rise, sea level rise faster, seals mating more, sewer bills rise, severe thunderstorms, sex change, sharks booming, sharks moving north, sheep shrink, shop closures, shrimp sex problems, shrinking ponds, shrinking shrine, ski resorts threatened, slow death, smaller brains, smog, snowfall increase, snowfall heavy, snowfall reduction, soaring food prices, societal collapse, songbirds change eating habits, sour grapes, space problem, spiders invade Scotland, squid population explosion, squirrels reproduce earlier, spectacular orchids, storms wetter, stormwater drains stressed, street crime to increase, suicide, Tabasco tragedy, taxes, tectonic plate movement, teenage drinking, terrorism, threat to peace, ticks move northward (Sweden), tides rise, tourism increase, trade barriers, trade winds weakened, tree beetle attacks, tree foliage increase (UK), tree growth slowed, trees could return to Antarctic, trees in trouble, trees less colourful, trees more colourful, trees lush, tropics expansion, tropopause raised, tsunamis, turtles crash, turtles lay earlier, UK coastal impact, UK Katrina, Vampire moths, Venice flooded, volcanic eruptions, walrus displaced, walrus pups orphaned, walrus stampede, war, wars over water, wars sparked, wars threaten billions, water bills double, water supply unreliability, water scarcity (20% of increase), water stress, weather out of its mind, weather patterns awry, weeds, Western aid cancelled out, West Nile fever, whales move north, wheat yields crushed in Australia, white Christmas dream ends, wildfires, wind shift, wind reduced, wine - harm to Australian industry, wine industry damage (California), wine industry disaster (US), wine - more English, wine -German boon, wine - no more French, winters in Britain colder, wolves eat more moose, wolves eat less, workers laid off, World bankruptcy, World in crisis, World in flames, Yellow fever. (Numbers Watch, U.K. 2007)

Urban Myth of Climate – An urban myth is a fictional tale or conspiracy theory that circulates widely, is told and retold and presumed true by mere repetition. Urban myths are spread and illuminated in mass media, and become part of popular culture. Urban myths circulate from no singular source and can survive as fearmongering news items. Myth-making eco-propagandists have become mainstream news media assets. Urban myths can be dangerous propaganda that causes civic alarm and paranoia often for nefarious political advantage. Examples are the global Y2K internet collapse, white privilege, Russian collusion, World War III and endless doomsday scenarios. For over a half century, environmentalists have menaced us with fear-mongering scenarios of man-made ecological apocalypse.

The environmental movement is the most densely-organized movement in human history. There are over 15,000 environmental nonprofits in the U.S. Having run out of green campaign grievances for their tax-exempt fundraising, radical eco-groups have come up with the most fantastic, intractable and theoretical problem of all – controlling dynamic, natural global climate variabilities (atmospheric thermodynamic radiative forcing). In scapegoating climate change for every inexplicable natural phenomenon and natural disaster, green-government tyranny would subvert capitalism itself with "climate austerity." Partisan green-government regulations inflate the costs of goods, services, activities, food, and energies.

CHAPTER 5

▼

CLIMATE CHANGE SCIENCE

"Science is a process for understanding how nature works, whereby we explore new ideas to find new representations of the world the explain what is observed. Part of science is to conduct experiments, make observations, do calculations and make predictions. But another part of science asks deep questions about how nature works.

What constitutes evidence in climate science? Scientific evidence is generally regarded to consist of observations and experimental results. In complex natural systems, the epistemic status of observation is not straightforward. There are data homogenization adjustments, model assimilation of observations, retrieval algorithms to interpret voltages measured by satellites and paleoclimate proxies. As a result, many climate data records are not without controversy, and there are ongoing revisions to many data records. Scientific investigations of the dynamics of the climate system have more in common with systems of biology and economics than physics and chemistry, owing to the inherent complexity of the system and the inability to conduct controlled experiments. Complexity is not the same thing as complicated. Complicated systems have many parts but simple chains of causation. Complexity of the climate system arises from the chaotic behavior and nonlinearity of the equations for motions in the atmosphere and oceans, and the feedbacks between subsystems for the atmosphere, oceans, land surface and glacial ice.

Climate change associated with increasing concentrations of atmospheric carbon dioxide (CO_2) is a theory in which the basic mechanism is well understood, but whose magnitude is highly uncertain. Here are the incontrovertible facts about global warming:

- Average global surface temperatures have overall increased since 1860,
- CO_2 has infrared emission spectra, and thus acts to warm the planet Earth,
- Humans have been adding CO_2 to the atmosphere via emissions from burning fossil fuels.

A changing climate has been the norm throughout the Earth's 4.3-billion -year history. The Earth's temperature and weather patterns change naturally over timescales ranging from decades to millions of years. Natural variations in surface climate originate in two ways:

1. Internal climate fluctuations associated with circulations in the atmosphere and oceans produce exchanges of energy, water and carbon between the atmosphere, oceans, land and ice.
2. External influences by humans are also changing atmospheric composition by increasing the emissions of carbon dioxide, and other greenhouse gases and by altering the concentrations of aerosol particles in the atmosphere.

The changing definitions by the UNFCCC currently define climate change as 'a climate change which is attributed directly or indirectly to *human activity* that alters the composition of the global atmosphere and which is in addition to natural climate variabilities observed over comparable time periods.'" Judith Curry, *Climate Uncertainty and Risk...*, Jan. 2023.

Here are some rational questions that should be asked about climate change.

1. What is "climate change"?

Climate change is a broad, general term largely the consequence of global warming. Climate change may include many natural phenomena such as the increasing incidence of extreme weather events like powerful hurricanes and severe drought to more frequent flooding and longer-lasting heat waves. And, may cause the accelerated ice melt in polar areas, and could be related to a rise in global sea levels. Climate impacts may worsen pollen seasons, spread vector-borne diseases, and have hundreds more suggested impacts.

2. What is causing climate change?

Climate change may be attributed to the increased levels of atmospheric greenhouse gases (GHGs) – chiefly carbon dioxide – produced by our burning of fossil fuels for electricity, industry, and transportation. Carbon dioxide, methane, nitrous oxide and other chlorofluohydrocarbon GHGs trap heat that would otherwise escape Earth's atmosphere (*atmospheric thermodynamic radiative forcing*). This is known as the "greenhouse effect," which in the right proportion, holds onto enough heat to support life on the planet Earth. Without them, the Earth would lose so much heat that life as we know it would be impossible. The problem arises when GHGs trap too much of the sun's energy as heat, upsetting the naturally-evolved systems that regulate our climate.

3. How could climate change be a problem?

If climate change causes more extreme weather, it could mean more damaged infrastructure, more lives and livelihoods lost to hurricanes, floods, and wildfires. It could mean more negative health outcomes, from heart and lung diseases linked to poor air quality to increased rates of vector-borne diseases, whose ranges are limited by temperature, expand as environmental conditions become more favorable. It could mean sea-level rise. More heat waves, drought, and erratic rainfall could impact basic food and water security, fuel migration, instability, and security concerns all over the globe.

4. Do scientists really agree on climate change?

In climate science there are "Believers" and "Skeptics." True scientists are trained skeptics in the policies and practices of the *Scientific Method*. People with greater scientific knowledge are more skeptical about the popular media horrors of global warming than those who have less scientific knowledge. Myth-making eco-propagandists have become mainstream news media assets. Today's gratuitous and politically-partisan green-government regulations inflate the costs of all goods, services, activities, food and energies, and kill jobs and prosperity – much the insidious result of the urban myth of anthropogenic climate change.

Believers are progressive prey of the partisan legacy media fearmongering game of climate hysteria and dangerous environmental industrial complex grifters (Gore, Kerry, etc.). Non-science Believers hold *climatism* as a religious rite and moral fetish. Believers support flawed fixes of suggested climate change without regard for the ultimate job-killing and global economic impacts of their tyrannical mandates to end fossil fuels.

Have no doubt that environmental awareness has benefitted both the human and natural environment with rigorous, science-based regulatory controls. However, eco-activism has spawned an "axis of antagonism" in militant and litigious eco-groups that spread fear-mongering propaganda the world over, at your personal and business expense. Expanded science literacy could mitigate the hysteria of eco-propaganda, and vastly improve science-based environmental policymaking.

No science has concluded that human activities are the proximate and dominate cause of global climate change (*Atmospheric Thermodynamic Radiative Forcing*). Though, there are legions of *climate warriors* that hold climate as a personal and professional fetish indulgence, and that will anxiously point to dozens incongruous computer models and hundreds of coincidental proxy natural phenomena as proof of their warming beliefs. In the absence of verifiable scientific evidence, it is the height of human conceit, arrogance and delusion to believe and advocate that humans can optimize global climates.

The global preoccupation (fetish) with climate alarmism has left essential conservation and pollution problems neglected in enforcement protections and government funding.

Believers vs. Skeptics

Believers – U.N. IPCC 2023 "Synthesis Report" states:

"Human activities, principally through emissions of greenhouse gases, have unequivocally caused global warming, with global surface temperature reaching 1.1°C above 1850-1900 in 2011-2020. Global greenhouse gas emissions have continued to increase, with unequal historical and ongoing contributions arising from unsustainable energy use, land use and land-use change, lifestyles and patterns of consumption and production across regions, between and within countries, and among individuals."

Skeptics – Science and Environmental Policy Project (SEPP) 2023 Review states:

"What we know about future and measured global average surface air temperature an about future global surface air temperature is nothing. The climate models cannot simulate present air temperature, cannot predict future air temperature, cannot resolve the effect of GHG emissions, cannot detect, attribute or project the impact, if any, of human fossil fuel emissions. The climate has probably warmed since 1900; the rate of warming is unknown. The magnitude of warming is unknown. There is no evidence of any unprecedented change. CO2 climatology lives on false precision. CO2 climatologists are not trained to evaluate the reliability of their own models and data. The UN IPCC claim of human-caused climate change has no basis in science."

Believers – Nature.com 2021 states:

"Prioritizing emission reductions neither equates to 'reduction only', nor does it mean delaying the ramp-up of carbon dioxide removal. Most modelled pathways to meet the Paris Agreement involve a significant scaling up of removals. Given that many important technologies are still in their infancy, much investment is and will be needed to ensure that there are enough removal options for residual emissions. We need to make progress as fast as realistically possible on both emission reductions and removals."

Skeptics – Nobel Laureate Dr. John Clauser Physicist 2023:

U.N. IPCC's 40 climate models do not produce consensus predictions of global temperatures. More importantly, IPCC models do not account for the dominant influence of global cloud cover that controls climate temperatures. Clouds (atmospheric water vapor) continuously cover 45% to 95% of the earth. Cloud behavior has evolved over millions of years to regulate our climates, and has a two-hundred times greater influence on global temperatures (atmospheric thermodynamic radiative forcing) as compared to the trace carbon dioxide or methane greenhouse gases.

Believers – NASA Announces 2023 Hottest Year, Sept. 2023:

"The summer of 2023 was Earth's hottest since global records began in 1880, according to scientists at NASA's Goddard Institute of Space Studies (GISS) in New York."

Skeptics – Science & Environmental Policy Project, Oct. 2023:

"In their Global Temperature Report, September 2023, John Christy and Roy Spencer discuss the unusual, sudden warming occurring in the atmosphere, where the greenhouse gases obstruct outgoing radiation. Contrary to many reports, there is no sudden increase in carbon dioxide concentrations. NOAA's Global Monitoring Laboratory for June 2023, the latest month available, reported the usual seasonal change with CO_2 concentrations increasing in the Northern Hemisphere winter and declining in the spring. In June 2023 CO_2 was at 419.51 ppm (parts per million volume) compared to June 2022 at 417.43 ppm. According to global warming alarmists, this slight change should have a significant impact on temperatures. It does not."

Believers – World Wildlife Fund, August 2023:

"Coral bleaching matters because once these corals die, reefs rarely come back. With few corals surviving, they struggle to reproduce, and entire reef ecosystems, on which people and wildlife depend, deteriorate. Bleaching also matters because it's not an isolated phenomenon.

According to the National Oceanic and Atmospheric Association, between 2014 and 2017 around 75% of the world's tropical coral reefs experienced heat-stress severe enough to trigger bleaching. For 30% of the world's reefs, that heat-stress was enough to kill coral."

Skeptics – Scripps Institute of Oceanography, July 2022:

"The largest global coral-bleaching event ever documented struck the world's oceans in 2014 and lasted until 2017. The onset of this abnormal whitening condition (caused by warmed sea and CO_2 acidification sea water) spawned widespread gloom-and-doom news reports about its calamitous effect on Australia's Great Barrier Reef and more general predictions of coral reef extinction by 2050. But a new 10-year study from Palmyra Atoll in the remote central Pacific Ocean shows that reefs outside the reach of local human impacts can recover from bleaching."

Believers – Forbes, July 2023:

"Between 2000 and 2019, annual deaths from heat exposure increased globally. The 20-year period coincided with the earth warmed by about 0.9 degrees Fahrenheit. The heat-related fatalities disproportionately impacted Asia, Africa and Southern parts of Europe and North America."

Skeptics – Forbes, July 2023:

"The NOAA's account of what it calls "weather-related deaths" suggests that during the 30-year period 1988 to 2017, an average of 134 heat-related deaths occurred annually, while 30 per year were cold-related. Contrary to the NOAA, the CDC's National Center for Health Statistics Compressed Mortality Database, which is based on actual death certificates, indicates that roughly twice as many people die of cold in a given year than of heat."

Believers – Pope Francis, *Laudate Deum* ('Praise God'), CFACT Oct. 2023:

"No one can ignore the fact that in recent years we have witnessed extreme weather phenomena, frequent periods of unusual heat, drought and other cries of protest on the part of the earth that are only a few palpable expressions of a silent disease that affects everyone."

Skeptics – H. Sterling Burnett, CFACT Oct. 2023:

"Focusing on the overall claim that extreme weather is getting worse over time, there is simply no data to support this claim. Trends of various extreme weather events do not show that they are occurring either more frequently or that such events are more severe when they do occur than they have been historically. The appearance of more frequent weather disasters around the globe is actually likely an artifact of new technology allowing 24-hour-a-day global online and broadcast media coverage, with the media hyping every bad weather event and tying it to climate change. The IPCC itself agrees with a study that came to the conclusion that "on the basis of observational data, the climate crisis that, according to many sources, we are experiencing today, is not evident yet."

Believers – United Nations, Feb, 2023:

"Transitioning to a Net Zero world is one of the greatest challenges humankind has faced. It calls for nothing less than a complete transformation of how we produce, consume, and move about. The energy sector is the source of around three-quarters of greenhouse gas emissions today and holds the key to averting the worst effects of climate change. Replacing polluting coal, gas and oil-fired power with energy from renewable sources, such as wind or solar, would dramatically reduce carbon emissions."

Skeptics – Science and Environmental Policy Project, Jan. 2023:

"Both Koonin and Nakamura understand climate modeling and expose deficiencies in establishment climate science and modeling procedures The issue of Net Zero carbon emissions continues to intensify as some citizens realize that the political leaders advocating Net Zero have no idea of the costs involved, and the costs continue to escalate. This is particularly true in the U.K. where it is becoming obvious that the policies are becoming disastrous, and no one thought the issues out.

Believers – John Kerry, U.S. Climate Ambassador, AP May 2023:

"Kerry said the deadline to watch is 2030. By then, the U.N.'s top climate panel says, the world will need to have nearly halved climate-damaging emissions to stave off the more devastating scenarios of global warming."

Skeptics – Paul Taylor, The Wall Street Journal 2022:

Urban Myth of Climate – An urban myth is a fictional tale or conspiracy theory that circulates widely, is told and retold and presumed true by mere repetition. Urban myths are spread and illuminated in mass media, and become part of popular culture. Urban myths circulate from no singular source and can survive as fearmongering news items. Myth-making eco-propagandists have become mainstream news media assets. Urban myths can be dangerous propaganda that causes civic alarm and paranoia often for nefarious political advantage. Examples are the global Y2K internet collapse, white privilege, Russian collusion, World War III and endless doomsday scenarios. For over a half century, environmentalists have menaced us with fear-mongering scenarios of man-made ecological apocalypse. The environmental movement is the most densely-organized movement in human history. There are over 15,000 environmental nonprofits in the U.S. Having run out of green campaign grievances for their tax-exempt fundraising, radical eco-groups have come up with the most fantastic, intractable and theoretical problem of all – controlling dynamic, natural global climate variabilities – Atmospheric Thermodynamic Radiative Forcing.

Believers – Environmental Defense Fund, 2023:

"If it seems as though the most intense hurricanes happen more often than they used to, you're right: The proportion of Atlantic Ocean hurricanes that are Category 3 or above has doubled since 1980. And if you're wondering how climate change has contributed, consider this: Over 90% of the heat trapped by greenhouse gases has been absorbed by the world's oceans. That means warmer waters, rising seas, higher wind speeds and more moisture in the atmosphere. These shifts are making hurricanes stronger, wetter and more likely to intensify rapidly, unleashing record-breaking downpours with little time for communities to evacuate."

Skeptics – Richard Lindzen Ph.D. Astrophysicist, Science and Environmental Policy Project, Sept. 2021:

"For about 33 years, many of us have been battling against climate hysteria. We have correctly noted: 1) The exaggerated sensitivity, 2) The role of other processes and natural internal variability, 3) The inconsistency with the paleoclimate record, and 4) The absence of evidence for increased extremes, droughts, flood, wildfires, and so on."

Believers – The Los Angeles Ecopolitics Examiner, 2014:

"Consistent with the myopic militancy of global green eco-groups, more than 100 worldwide protests by environmentalists are planned in an effort to stop fracking. The Global Frackdown website and campaign by the Washington, D.C. nonprofit Food & Water Watch claims that fracking "...has already damaged communities and ruined lives. It pollutes water and makes people sick. The message is, we need to ban fracking. We think fracking is just another way to produce dirty fossil fuels."

Skeptics – The Los Angeles Ecopolitics Examiner, 2014:

"The cause of U.S. CO2 reduction is the growing economical conversion from coal to natural gas in U.S. energy production. Natural gas emits 45% less CO2 than coal in energy production. Three decades of America's petroleum industry development of underground hydraulic fracturing technology (*fracking*) has resulted in massive reductions of U.S. CO2 emissions – about twice the total reduction compared to the Kyoto Protocol regulations for CO2 reductions in the E.U., and much of the developed economies. Fracking to replace coal with natural gas in energy production has succeeded where Kyoto regulations and carbon cap-and-trade taxes have failed."

CHAPTER 6

▼

ENERGY AND ECONOMY

Today, partisan eco-groups are the architects of many job-killing and unnecessary environmental regulations. The green lobby has become less covert about their true subversive intentions – which are the undoing of capitalism. For over a half century, environmentalists have menaced us with fear-mongering scenarios of man-made ecological apocalypse. Most of the developed world has cleaned up their environments after going through the growing pains of economic development and political maturity.

Having run out of green campaign grievances for their fundraising, radical eco-groups have invaded the U.N.'s IPCC to come up with the most fantastic, intractable and theoretical problem of all – controlling dynamic, natural global climate variabilities. In scapegoating "climate change" for every inexplicable natural phenomenon and natural disaster as gratuitous "green dogma," eco-socialists would subvert capitalism itself.

Climate science remains unsettled. The unpredictable interactions and ultimate atmospheric fates of clouds and aerosols stand in the way of reliable global warming cause-and-effect findings and realistic climate predictions. Neither government nor eco-alarmists nor technology can separate global greenhouse gas emissions from human security and prosperity. The world now spends about $500 billion per year on climate controls. Today, climate control costs are more predictable than the climate control benefits. Therefore, green government tyranny could become the existential threat to humanity, not climate.

Energy

A majority of Americans consider climate change a priority today so that future generations can have a sustainable planet, and this view is held across generations.

Looking to the future, the public is closely divided on what it will take to address climate change: While about half say it's likely major lifestyle changes in the U.S. will be needed to deal with climate change impacts, almost as many say it's more likely new developments in technology will address most of the problems cause by climate change.

On policy, majorities prioritize the use of renewable energy and back the expanded use of specific sources like wind and solar. Americans offer more support than opposition to a range of policies aimed at reducing the effects of climate change, including key climate-related aspects of President Joe Biden's recent infrastructure proposal. Still, Americans do not back a complete break with carbon: A majority says oil and gas should still be part of the energy mix in the U.S., and about half oppose phasing out gas-powered vehicles by 2035.

Overall, 64% of U.S. adults say reducing the effects of climate change needs to be "a top priority to ensure a sustainable planet for future generations, even if that means fewer resources for addressing other important problems today." By contrast, 34% say that reducing the effects of climate change needs to be "a lower priority, with so many other important problems facing Americans today, even if that means more climate problems for future generations."

There are stark partisan differences over this sentiment. Nearly nine-in-ten Democrats (87%) say efforts to reduce the effects of climate change need to be prioritized today to ensure a sustainable planet. By contrast, 61% of Republicans say that efforts to reduce the effects of climate change need to be a lower priority, with so many other important problems facing Americans today. (Democrats and Republicans include those who lean to each party.)

Asked to look to the future 50 years from now, 51% of Americans say it's more likely that major changes to everyday life in the U.S. will be needed to address the problems caused by global climate change. By contrast, 46% say it's more likely that new technology will be able to address most of the problems caused by global climate change.

Most Democrats (69%) expect that in 50 years major lifestyle changes in the U.S. will be needed to address the problems caused by climate change. By contrast, among Republicans, two-thirds (66%) say it's likelier that new technology will be able to address most climate change problems in the U.S. among Republicans, this view is widely held (81%) among the majority who do not see climate change as an important personal concern; Republicans who express greater personal concern about climate change are more likely to say major changes to everyday life in the future will be needed to address problems caused by climate change. Overall, majorities across generations believe that climate change should be a top priority today to ensure a sustainable planet for future generations. Generational divisions are more prominent among Republicans than Democrats.

Among Republicans, about half of Gen Zers (49%) and Millennials (48%) give top priority to reducing the effect of climate change today, even if that means fewer resources to deal with other important problems. By contrast, majorities of Gen X (61%) and Baby Boomer and older Republicans (71%) say reducing the effects of climate change needs to a lower priority today, given the other problems Americans are facing. Generational differences among Democrats on this question are modest, with clear majorities giving priority to dealing with climate change today. (Pew Research, May 2021)

The world is waking up to the fact that the climate policy goal of achieving "Net-Zero" CO_2 emissions brings crippling economic pain. Fossil fuel prices shot up by 26 per cent across industrialized economies last year and will rise globally by another 50 per cent this year. Politicians blame Russia's invasion of Ukraine, but the long-term trend stems mostly from governments demonizing fossil fuels while their societies remain dependent on them. Since the 2015 Paris Accords climate agreement, global investment in fossil fuels has halved, inevitably driving up prices.

As fossil fuel prices climb, activists believe people will shift painlessly to renewable energy sources. But they've made a major miscalculation: renewables are far from ready to power the world. Solar and wind can only work with massive amounts of backup power, mostly fossil fuels, to keep the world running when the wind dies down, the sky clouds over, or night falls. Moreover, renewables mostly generate electricity, which is just one-fifth of our total energy use — the vast majority is non-electric like transport, industrial processes and heat.

That's why the world still gets 80 per cent of its energy from fossil fuels, and renewables deliver just 15 per cent. There won't be change any time soon — even the Biden administration expects the world in 2050 to be dependent on fossil fuels for 70 per cent of its energy.

But most "Net-Zero" policies try to force much greater reductions in fossil fuels, driving down investments and making them extremely expensive before alternatives can take over. That leads to worldwide pain, like the Northern Hemisphere winter we're entering where Europe prepares for brownouts and two-thirds of the U.K. population is predicted to enter energy poverty.

Rich countries are showcasing the policies to avoid. Germany is on track to spend more than half a trillion dollars on climate policies by 2025, yet has only managed to reduce fossil fuel dependency from 84 per cent in 2000 to 77 per cent today. McKinsey estimates that getting to Net-Zero will cost Europe 5.3 per cent of its GDP in low-emission assets every year, or more than U.S. $200 billion annually just for Germany. That is more than it spends annually on education and police, courts and prisons combined. (Bjorn Lomborg, National Post Oct, 2022)

The developed world's response to the global energy crisis has put its hypocritical attitude toward fossil fuels on display. Wealthy countries admonish developing ones to use renewable energy. Last month the "Group of Seven" went so far as to announce they would no longer fund fossil-fuel development abroad. Meanwhile, Europe and the U.S. are begging Arab nations to expand oil production. Germany is reopening coal power plants, and Spain and Italy are spending big on African gas production. So many European countries have asked Botswana to mine more coal that the nation will more than double its exports.

The developed world became wealthy through the pervasive use of fossil fuels, which still overwhelmingly power most of its economies. Solar and wind power aren't reliable, simply because there are nights, clouds and still days. Improving battery storage won't help much: There are enough batteries in the world today only to power global average electricity consumption for 75 seconds. Even though the supply is being scaled up rapidly, by 2030 the world's batteries would still cover less than 11 minutes. Every German winter, when solar output is at its minimum, there is near-zero wind energy available for at least five days—or more than 7,000 minutes.

This is why solar panels and wind turbines can't deliver most of the energy for industrializing poor countries. Factories can't stop and start with the wind; steel and fertilizer production are dependent on coal and gas; and most solar and wind power simply can't deliver the power necessary to run the water pumps, tractors, and machines that lift people out of poverty.

That's why fossil fuels still provide more than three-fourths of wealthy countries' energy, while solar and wind deliver less than 3%. An average person in the developed world uses more fossil-fuel-generated energy every day than all the energy used by 23 poor Africans.

Yet the world's rich are trying to choke off funding for new fossil fuels in developing countries. An estimated 3.5 billion of the world's poorest people have no reliable access to electricity. Rather than give them access to the tools that have helped rich nations develop, wealthy countries blithely instruct developing nations to skip coal, gas and oil, and go straight to a green nirvana of solar panels and wind turbines. (Bjorn Lomborg The Wall Street Journal, June 2022)

Economics

Economic Systems: "Classical economic systems could be separated by two questions. One, who controls the means of production (farms, factories, etc.); and two, who controls what is produced. In a free-market system, individuals, and groups of individuals (corporations) control the means of production; and markets (individuals acting as groups) control what is produced.

In Communism, government controls both the means of production and what is produced.

In Socialism, government (no matter which type) controls the means of production and what is produced.

In Fascism, government controls what is produced, but not the means of production.

Many advocates of the Green New Deal are advocating moving towards a Fascist system – favoring certain types of energy for electricity generation such as wind and solar while punishing other types of energy generations such as fossil fuels or nuclear." (Science and Environmental Policy Project, Haapala July 2022)

Environmental economics is the study of the cost-effective allocation, use, and protection of the world's natural resources.

Economics, broadly speaking, is the study of how humans produce and consume goods and services. Environmental economics focuses on how they use and manage finite resources in a manner that serves the population while meeting concerns about environmental impact. This helps governments weigh the pros and cons of alternative measures and design appropriate environmental policies. Environmental economics studies the impact of environmental policies and devises solutions to problems resulting from them.

Key Principles

- Environmental economics can either be prescriptive-based or incentive-based.
- A major subject of environmental economics is externalities, the additional costs of doing business that are not paid by the business or its consumers.

- Another major subject of environmental economics is placing a value on public goods, such as clean air, and calculating the costs of losing those goods.
- Since some environmental goods are not limited to a single country, environmental economics often requires a transnational approach.

The basic theory underpinning environmental economics is that environmental amenities (or environmental goods) have economic value and there are costs to economic growth that are not accounted for in more traditional models.

Environmental goods include things like access to clean water, clean air, the survival of wildlife, and the general climate. Although it is hard to put a price tag on environmental goods, there may be a high cost when they are lost. Environmental goods are usually difficult to fully privatize and subject to the *tragedy of the commons*.

Destruction or overuse of environmental goods, like pollution and other kinds of environmental degradation, can represent a form of market failure because it imposes negative externalities. Environmental economists analyze the costs and benefits of specific economic policies that seek to correct such problems, and they may run theoretical tests or studies on the possible consequences of these policies.

Environmental economists are concerned with identifying specific problems, but there can be many approaches to solving the same environmental issue. If a state is trying to impose a transition to clean energy, for example, they have several options. The government can impose a fixed limit on carbon emissions, or it can adopt more incentive-based solutions, like placing quantity-based taxes on emissions or offering tax credits to companies that adopt renewable power sources.

All of these strategies rely on state intervention in the market, but some governments prefer to use a light touch and others may be more assertive. The degree of acceptable state intervention is an important political factor in determining environmental economic policy.

Broadly speaking, environmental economics may produce two types of policies:

1. Prescriptive Regulations

   In a prescriptive approach, the government dictates specific measures to reduce environmental harm. For example, they may prohibit highly-polluting industries, or require certain emissions-controlling technologies.

2. Market-Based Regulations

   Market-based policies use economic incentives to encourage desired behaviors. For example, cap-and-trade regulations do not prohibit companies from pollution, but they place a financial burden on those who do. These incentives reward companies for reducing their emissions, without dictating the method they use to do so.

Challenges of Environmental Economics

Because the nature and economic value of environmental goods often transcend national boundaries, environmental economics frequently requires a transnational approach. For example, an environmental economist could identify overfishing as a negative externality to be addressed.

The United States could impose regulations on its own fishing industry, but the problem wouldn't be solved without similar action from many other nations. The global character of such environmental issues has led to the rise of non-governmental organizations (NGOs, or nonprofits) like the Intergovernmental Panel on Climate Change (IPCC), which organizes annual forums for heads of state to negotiate international environmental policies.

Another challenge of environmental economics is the degree to which its findings affect other industries. More often than not, findings from environmental economists can result in controversy, and their policy prescriptions may be difficult to implement due to the complexity of the world market.

The presence of multiple marketplaces for carbon credits is an example of the chaotic transnational implementation of ideas stemming from environmental economics. Fuel economy standards set by the Environmental Protection Agency (EPA) are another example of the balancing act required by policy proposals related to environmental economics.

In the U.S., policy proposals stemming from environmental economics tend to cause contentious political debate. Leaders rarely agree about the degree of externalized environmental costs, making it difficult to craft substantive environmental policies. The EPA uses environmental economists to conduct analysis-related policy proposals.

These proposals are then vetted and evaluated by legislative bodies. The EPA oversees a National Center for Environmental Economics, which emphasizes market-based solutions like *cap-and-trade* policies for carbon emissions. Their priority policy issues are encouraging biofuel use, analyzing the costs of climate change, and addressing waste and pollution problems. (Investopedia, June 2023)

A McKinsey & Co. 2023 study finds that Net-Zero will cost a family in U.S. close to $20,000 every year. (Bjorn Lomborg, March 2023)

For over 50 years local, state and international environmental regulations have embedded cost increases in all of our goods, services, foods, energies and activities; resulting, in growing instances, in *climate austerity*. Accordingly, there should be a pause to reset government environmental regulations, and a re-alignment of the *Scientific Method* with environmental and economic policies.

Epilogue

Enrich your life with a boundless curiosity and respect for the Earth's biological and physical environmental wonders.

Paul Taylor

Bibliography and References

American Council on Science and Health, July 1999

Bjorn Lomborg, Mar. 2023

Bjorn Lomborg, National Post Oct. 2022

Bjorn Lomborg, The Wall Street Journal June 2022

Bjorn Lomborg, Copenhagen Consensus March 2023

CFACT, Aug. 2023

"Climate Uncertainty & Risk" Judith Curry, Jan. 2023

"Climate of Ecopolitics – A Citizens Guide" Paul Taylor, 2008

Eco America Website, Jan. 2008

Environmental Defense Fund, 2023

"Extraordinary Costs Of Green Energy..." — Manhattan Contrarian

Fact Sheet – Biden Climate Action..., April 2023

Gilder Technology Report, May 1999

H. Sterling Burnett, CFACT Oct, 2023

Hudson Institute, April 1999

John Clausner Ph.D., July 2023

Los Angeles Ecopolitics Examiner, 2014

NASA Hottest Year News, Sept. 2023

New York Post, Oct. 2023

Numbers Watch, U.K. Dec. 2007

Pew Research, May 2021

Pope Francis Praise Lord, CFACT Oct. 2023

"Science and Environmental Policy Project," Haapala July 2022

Scripps Institute of Oceanography, July 2022

The Wall Street Journal, Aug. 1999

U.N. IPCC Synthesis Report, 2023

U.N. on Climate, Feb. 2023

World Wildlife Fund, Aug. 2023

www.climate.nasa.gov/faq/34/what-types-of-data-do-scientists-use-to-study-climate/

www.commentary.org/noah-rothman /

www.wikipedia.org/wiki/Bible Prophecy

"15 Historical Nonprofit Moments You Should Know" - Whole Whale

"26 Incredible Nonprofit Statistics [2023]: How Many Nonprofits Are In The U.S.?a